PETERSHAM: RADAR AND OPERATIONAL RESEARCH

1940-1946

J M LEE

Richmond Local History Society..

Petersham: radar and operational research 1940-1946

By J M Lee

Richmond Local History Society

First published 2011

Reprinted 2016

Second edition 2024

Published in January 2024 by the Richmond Local History Society,
9 Bridge Road, St Margarets, Twickenham TW1 1RE

Edited by Robert Smith and Simon Fowler

Produced by Robert Smith

Designed by Mary Pollard

Printed by imprintdigital.com, Seychelles Farm, Upton Pyne, Exeter,
Devon EX5 5HY

ISBN 978-1-912314-04-1

Contents

Michael Lee 1932-2022 4

Introduction to the second edition 5

1: The transformation of Petersham 6

2: The Air Defence Research Group and the AA Radio School 10

3: The value of civilian scientists 15

4: Anti-aircraft radar 19

5: The purpose of the Petersham unit 23

6: The initial set up 27

7: Subsequent developments 33

8: The needs of the Army 40

9: Petersham's legacy 43

Sources 45

Appendix 1 48

The principal scientists working in Petersham 1940-1946

Appendix 2 49

Petersham as a birthplace of radio astronomy

by Timothy M M Baker, 2024

Michael Lee 1932-2022

JOHN MICHAEL LEE passed away on 2 February 2022, only a few weeks before what would have been his 90th birthday.

Born in Sheffield on 29 March 1932, he was educated at secondary school in Nottingham and at Christ Church, Oxford.

Michael's lifelong passion for local history started whilst he was still at school, writing and publishing a local history of Castle Donington where he lived. On graduation, he worked on the Victoria County History of Leicestershire and began an academic career as a lecturer in government at Manchester University, following two years on secondment to the civil service. He then became senior lecturer and reader at the Institute of Commonwealth Studies and Birkbeck College. His final academic post, from which he retired as emeritus professor in 1992, was chair of the politics department at the University of Bristol and dean of the Social Science faculty.

During his career he published numerous books, always well researched with political and social comment. The topics ranged from *Social Leaders and Public Persons* (Michael's book on local government in Cheshire, published by Oxford University Press) to *African Armies and Civil Order* (published by Chatto and Windus).

He and his wife Joy lived between Ealing and Bristol for 20 years. In 1998 they moved to Cross Deep in Twickenham where Michael took a keen interest in the area's local history. His thoroughly researched 168-page book *The Making of Modern Twickenham* was published by Historical Publications in 2005.

As chair of its publications sub-committee, Michael played a key role in expanding the Richmond Local History Society's publications programme, and his academic area of interest led to his researching the part that Petersham had played in Britain's wartime radar operations. In 2011 the Society published his book *Petersham: radar and operational research 1940-1946.* He co-wrote, with the late Leonard Chave, the Society's book *Ham and Petersham in Wartime*, also published in 2011, and several articles for *Richmond History*, the Society's journal, which he edited from 2002 to 2005.

Introduction to the second edition

THE IMPORTANT CONTRIBUTION OF BRITISH RADAR to the Allies' victory in the Second World War owes much to the secret experiments that were conducted in the village of Petersham (then in Surrey, and now in the London Borough of Richmond upon Thames).

Petersham was transformed during the Battle of Britain in September 1940 when Anti-Aircraft Command took over several buildings, including All Saints' Church and the vicarage, for its work and then set up radar transmitters on the golf course at Sudbrook Park.

Michael Lee's account in 2011 of Petersham's unique wartime role has been of continuing interest, not just to those interested in our local history but also to people much further afield.

In this second edition, we have re-presented Michael Lee's text, with some minor additions and amendments, in a new format that we hope will also appeal to an even wider audience. In doing so, we have also drawn on Michael's contributions to the Richmond Local History Society's 2011 book *Ham and Petersham in Wartime* and his article "Petersham at War", published in the 2007 issue of the Society's journal *Richmond History*.

This new edition also includes additional material by Timothy M M Baker on Petersham as a birthplace of radio astronomy.

Robert Smith and Simon Fowler
Richmond Local History Society
January 2024

1: The transformation of Petersham

All Saints' Church, Petersham
Wikimedia Commons / John Salmon

AFTER A VERY DRY SUMMER, Petersham was transformed in September 1940 during the Battle of Britain when Anti-Aircraft Command requisitioned the village hall and institute, the vicarage[1] and All Saints' Church[2] and then set up radar transmitters on the golf course. Elm Lodge, Cecil House and Cornerways were commandeered a little later, and huts were erected on the vicarage garden's lawns.

The village at that point of the hostilities contained fewer than 900 inhabitants and a number of evacuees. Pembroke Lodge in Richmond Park had at the same time just been taken over as the officers' mess of the GHQ Reconnaissance Unit, known as the Phantom Regiment, which also commandeered Richmond Hill Hotel to be its headquarters.

It was a tense moment for the whole country. The government had issued its "invasion imminent" signal to the armed forces on 7 September. The German Luftwaffe made a series of major bombing raids on London on the

[1] Now known as the Old Vicarage: Petersham acquired a new vicarage in 1997.

[2] All Saints' Church is now a private house.

9th, 11th and 13th in an attempt to achieve the air superiority necessary for German invading land forces.[3]

The most stressful day was the 15th. The Royal Air Force was sufficiently successful in the defence of British shores for Hitler on 17 September to postpone "Operation Sea Lion", his planned invasion of England.

Civilians had very limited access to Richmond Park. The Pioneer Corps guarded its gates; civil defence teams undertook experiments in its grounds; various sets of troops used its assault courses; and fires were frequently lit within its boundaries at night as decoys to deceive German aircraft into unloading their bombs onto open ground and not on London's buildings. These decoys (codenamed "Starfish") had large quantities of oil, tin foil and various devices to mimic sparks and poorly blacked-out windows. Richmond's set was moved to Hampstead Heath in May 1942.

The park contained two anti-aircraft gun batteries, ZS19 and ZS20. Each battery had its own bunker, plotting room and observatory. In what is now Greater London there were 78 similar batteries with around 200 guns. All anti-aircraft and coastal batteries had been on full alert since 24 August 1939 when it looked as if Hitler was about to invade Poland. Anti-Aircraft (AA) Command relied heavily on the 3.7-inch gun in fixed gun-sites, not on mobile batteries.

In Richmond Park, fires were frequently lit at night as decoys (codenamed 'Starfish') to deceive German aircraft into unloading their bombs onto open ground and not on London's buildings

Village life therefore became very militarised. Locals had to get used not only to providing billets for soldiers but also to the comings and goings of scout cars, motor bikes, lorries and the buses which brought in those who had been placed in Richmond boarding houses. They might find themselves bumping into young officers from the Phantom Regiment running for exercise during the night, because their commanding officer had ordered them to develop different sleep patterns so that they could be posted to any

[3] Richmond was first attacked in the early hours of 9 September, when high explosive bombs fell on houses in Mount Ararat Road and on Marchmont Road on Richmond Hill. There were no casualties. See Simon Fowler *Richmond at War 1939-1945*, p.14, Richmond Local History Society, 2015.

front line at a moment's notice. These young men were trained as observers who could report back to the field commander either on foot or by bike.

Stephen Paull, a physics teacher from Michigan, who was a trainee at Petersham between November 1941 and February 1942, wrote in his diary: "I visited Richmond Park accompanied by a British signal officer. The park is fenced off with barbed wire and guarded. Inside the park is kept all kinds of army radio equipment. The section of Richmond near the park appears to be an ordinary residential section of old houses and apartment hotels. To look at them would reveal nothing. But actually they are radio schools, offices, workshops, troop barracks and radio control positions."[4]

The village at a later stage had to accept the regular movement of caravans which carried the staff sent to monitor both guns and air raids. The radar receivers tested on trailers and the three-ton lorries used as recording vans were hardly suitable for Sudbrook Lane.

But the military presence had a distinctive tone. A high proportion of the men sent to Petersham were university graduates and even university teachers. The village was often visited by many distinguished scientists who conducted seminars and then retired for a drink at the Fox & Duck or The Dysart Arms. A standard joke was to comment on the density of "future Fellows of the Royal Society behind every corner".[5] Experts were always on the move; the armed forces often competed for their attention.

The vicarage[6] had the feel of something like a Cambridge college common room. (There was only one prominent Oxford don.) Connections made through scholarship or just college life were often more important to the promotion of research than the formalities of government. Those scientists passing through Petersham had often attended each other's lectures, examined each other's theses or read each other's learned papers. Many knew each other through either the Cavendish Laboratory in Cambridge or the "Tots and Quots", a private dining club started by Solly Zuckerman in 1931 and revived in 1939.

[4] Entry for 4 December 1941. Stephen Paull Papers, Imperial War Museum

[5] See Appendix 1 (p.48) for the principal scientists who came to Petersham, many of whom became Fellows of the Royal Society.

[6] The vicar, a bachelor who moved to live in the Old Stables on River Lane, continued to serve the parish from St Peter's Church.

John Cockcroft in 1951
Wikimedia Commons/Dutch National Archives

The Anti-Aircraft Command establishment in Petersham was closely linked to the Air Defence Research and Development Establishment (ADRDE) in Christchurch, Hampshire, supervised by John Cockcroft (1897-1967).

Cockcroft was a distinguished nuclear physicist from Cambridge who had recruited and led the teams that set up the first set of coastal radar stations. He then become an assistant director of research for the Ministry of Supply. Whenever he stayed in London, Cockcroft, a frequent visitor to Petersham, regularly used 28 Tufton Court, Westminster, the flat that belonged to Patrick Blackett who directed the Petersham establishment. This habit symbolised the intimacies of the whole enterprise

2: The Air Defence Research Group and the AA Radio School

THE AIR DEFENCE RESEARCH GROUP had begun work before being moved to Petersham.[7] For part of July and the whole of August 1940 it occupied first Savoy Hill House and then Brettenham House near Waterloo Bridge.

It is not clear why Petersham was chosen for its location. Perhaps those serving in the anti-aircraft batteries in Richmond Park saw the advantages of All Saints' Church and the village institute as large covered halls in which equipment could be hidden. There was value in placing research scientists close to gun batteries in action.

3.7-inch anti-aircraft gun in Richmond Park, 1940. ©*IWM*

The move to Petersham also gave Anti-Aircraft Command the space to set up a radio school which was an essential part of the operation. There had already been training courses for officers manning the coastal radar stations. The Petersham school was to train officers who could then be sent to advise gun batteries across the whole country. Petersham became a "radar university" with research and training in close proximity.

The whole establishment was a direct consequence of the discussions that followed the return on 13 June 1940 of Professor A V Hill (1886-1977) from the United States, where he had been exploring the possibility of sharing British scientific secrets with the American authorities.

[7] It was presumably moved to avoid The Blitz. In the early stages of the war, scientific bodies were transferred from London to safer places in other parts of the country.

Professor Hill combined many roles. He was secretary of the Royal Society, an independent MP for Cambridge University, the author of textbooks on anti-aircraft gunnery, a leading physiologist specialising in biophysics, and an expert on the heat generated in muscles.

Petersham became a 'radar university' with research and training in close proximity

Hill advised General Sir Frederick Alfred Pile (1884-1976), known as "Tim", the Commanding Officer of Anti-Aircraft Command, that he should appoint as his scientific adviser Professor Patrick Blackett (1897-1974).

Blackett was a professor of physics at the University of Manchester who was then on secondment to the instrument section of the Royal Aircraft Establishment working on the improvement of bomb-sights. His team there produced what was known as Mark XIV.

The "Research Group", formed by Blackett with Hill's help, was known affectionately as "Blackett's Circus". Blackett described it as "one of the first groups to be given the facilities for the study of a wide range of operational problems, the freedom to seek out these problems on their own initiative, and sufficiently close contact with the Service operational staffs to enable them to do this".[8]

The first members of the team nominated by Professor Hill were all physiologists, his own discipline. They included his 25-year-old son David, his son's friend Andrew Huxley (aged 23), and Leonard Bayliss, a lecturer in physiology at University College, London, where Bayliss's father, Sir William Bayliss, had promoted this particular scientific specialism.

Andrew Porter (aged 30), meeting Blackett who was then on an inspection with a government committee of the National Physical Laboratory in Teddington, asked for a transfer to more interesting work and became the fourth recruit. Porter almost immediately recommended a transfer from Army service to the Petersham training school of his brother, Ronald, and a school friend from Ulverston, Harry Tyson. Ronald Porter performed so well on the course and in his posting to a gun battery that he became a scientific adviser to General Pile.

[8] P M S Blackett *Studies of War: nuclear and conventional*, p.206, Oliver & Boyd, 1962

The first principal of the "radio school" opened in October 1940 was equally distinguished. Jack Ratcliffe (1902-1987) had been the head of the radio ionosphere research group in the Cavendish Laboratory of Cambridge University and was therefore an expert on the reflection of radio waves. At the outbreak of war he had joined the Air Ministry Research Establishment, heading the group using a new type of ground radar equipment.

Ratcliffe took over Elm Lodge as his headquarters after Blackett had lodged his staff in the vicarage. They both shared the space in All Saints' Church and the village institute. The church (1909) and the institute (1901) – gifts to the parish from Mrs Warde, who built them as memorials to her father and her aunt – became covers for gun-laying radar set testing and for gun battery training exercises.

All Saints' Church and the village institute became covers for gun-laying radar set testing and for gun battery training exercises

Hill, Blackett and General Pile, in setting up the Petersham establishment, were not acting simply on behalf of Anti-Aircraft Command. They were all part of a wider set of secret discussions instigated in 1935 by the Air Ministry which had brought civilian scientists into preparations for war, undertaken while the government pursued a policy of appeasement towards the Hitler regime in Germany. They were familiar with the progress already made in research and development on radar, and with the rivalry between the three armed services, each of which saw its potential within a framework of different strategic priorities.

Anti-Aircraft Command had been created in April 1939 as a way of placing the AA divisions of the Army under the operational control of RAF Fighter Command. It stood somewhat awkwardly between the Army and the Royal Air Force and required certain skills in inter-service diplomacy in order to be heard in government.

The headquarters of Anti-Aircraft Command was deliberately placed in Glenthorn, near Stanmore in Middlesex, so that it was close to RAF Fighter Command headquarters at Bentley Priory. General Pile had been the major-general in charge of the 1st Anti-Aircraft Division guarding London. He had a very good relationship with Sir Hugh Dowding, the head of Fighter

Command, and an enthusiasm that set the tone of their collaboration. The members of the Petersham staff were all familiar with Stanmore.

Hill and Blackett, like Sir Henry Tizard with whom they were closely associated and General Pile whom they advised, had been in the armed forces during the First World War. Hill in the Cambridgeshire Regiment directed a section of the munitions development department which experimented in anti-aircraft gunnery; Blackett was in the Navy at the battle of Jutland; Tizard was an experimental officer in both the Artillery and the Royal Flying Corps; Pile had also been in the Artillery, and after the war was a regular in the Royal Tank Corps. They understood each other.

Hill and Blackett were fully aware of the extremely secret MAUD committee set up in the spring of 1940 to advise government on the feasibility of an atomic bomb. Both had been involved in arguments about the role of scientists in politics. Blackett, having resigned from the Navy and read for a degree in mathematics and physics, entered the Cavendish Laboratory as a research student in 1921, and with an interest in cosmic rays had risen to have his own laboratory in Birkbeck College by 1933. He had regular holidays with other of the left-wing intellectuals who enjoyed the hospitality of Clough Williams-Ellis at Portmeirion, and participated in the debates on the "social responsibility" of science. Hill found it convenient not to antagonise these "radicals" because he saw the importance of recruiting their services for war. The Penguin paperback *Science in War,* published in 1940, summarised the value of their enterprise.

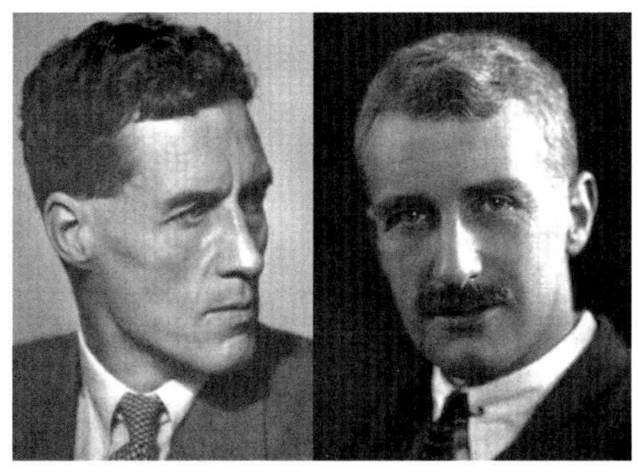

Patrick Blackett (left) and Archibald Vivian Hill
The Nobel Foundation/ Wikimedia Commons

Jack Ratcliffe, the first principal of Petersham's "radio school", photographed in 1966.
© National Portrait Gallery, London

Hill and Blackett were also conversant with the tension generated in discussions about the application of radar to anti-aircraft guns. Frederick Lindemann (1886-1957) (later Lord Cherwell), the adviser to Winston Churchill who was opposing the policy of appeasement, had regular disagreements with Sir Henry Tizard, the Rector of Imperial College, who was the Air Ministry's principal adviser before war was declared. Hill and Blackett supported Tizard on many occasions.

The collapse of the Chamberlain government in May 1940 and the massive presence of Churchill as the new prime minister of a coalition changed the tone of the debate. Hill played an important role in the promotion of Tizard's plan to secure American assistance in radar development; he was in Washington DC at the time of the crisis of confidence in Chamberlain. The project to send a scientific mission to the United States, hatched while Chamberlain was still in office, was approved and implemented by Churchill. Tizard was appointed to lead the mission.

Blackett began work with the Air Defence Research Group knowing very well that the creation of his team was a gamble which would only be successful if the existing collaboration between civilian scientists and military personnel gained effective government backing.

3: The value of civilian scientists

BOTH HILL AND BLACKETT had been prominent members of the Air Ministry Committee for the Scientific Survey of Air Defence (CSSAD) set up in 1935 and chaired by Sir Henry Tizard.

Sir Henry Tizard
©National Portrait Gallery, London

This committee, whose secretary was A P Rowe (known as "Jimmy"), had pressed for the recruitment of civilian scientists to develop new technologies. Rowe himself, a lecturer at Imperial College, had been taken into the Air Ministry to be a personal assistant to H E Wimperis, the department's director of scientific research, who had previously supervised the Air Ministry laboratory in the same college.

Frederick Lindemann, Winston Churchill's scientific adviser, became a member of its technical sub-committee. Lindemann had been a close friend of Tizard before the First World War, but the two began to quarrel after Lindemann suspected that the committee had been formed to undermine some of his own proposals. Growing animosity between the two men became part of the discussion on strategic questions after Churchill as prime minister continued to rely on Lindemann's judgement.

All members of "the Tizard Committee" knew that existing observation posts and sound locators gave inadequate warnings of approaching enemy aircraft. Professor Hill, an expert on anti-aircraft guns, had also written a book on sound locators. It was imperative to create a new system of command and control.

In "the war of the radio waves" the military planners of grand strategy had to improvise by using civilians for political reasons.

In the first place, hired scientists could make secret preparations fairly cheaply and without fuss during the period of appeasement. Until the Munich crisis of September 1938, there was no easy way of guessing when war would break out.

In the second place, the prospect of the whole civilian population being vulnerable to enemy bombing made air raid precautions into a matter of collaboration between different branches of government. Hired scientists could begin to make the necessary joined-up thinking by spelling out the alternatives to the authorities. Papers written by experts for the 1939 conference on air-raid shelters, for instance, were classified as secret. Protection against air attack was naturally a sensitive political issue.

Nazi Germany had no such prevarications. Its preparations for war were well advanced. Its scientists and engineers were directly engaged on weapons development projects. Germany's air force, the Luftwaffe, had control of its own anti-aircraft guns and searchlights; the Royal Air Force in Britain had to collaborate with the Army on civil defence questions through Anti-Aircraft Command.

The invention of radar presented a challenge to the conduct of operations at all levels. The Petersham establishment had to build the right connections between the Royal Air Force and the Army – between radar research and the procurement of army equipment.

There appeared to be a sharp contrast between RAF and Army attitudes to science. The RAF seemed to encourage experimentation in ways which the Army did not. As the war progressed, Anti-Aircraft Command developed the reputation of being a state within a state, having such good personal relationships that it could secure privileges outside the normal channels of bureaucracy.

Hill had long been keen to work with the Americans, an idea first officially put forward in May 1939. There were difficulties on both sides. But, in March 1940, Hill was sent to Washington DC, technically as an air attaché in the British Embassy but in practice as the chief negotiator for some kind of British scientific mission. Hill, after a ten-week visit, proposed "a frank offer

MI6's principal scientist, who lived in a flat on Richmond Hill and was thoroughly familiar with Petersham and its wartime role, was able to identify the radio beams used to guide German bombers to their targets

to exchange information and experience"; he had the full backing of Lord Lothian, the British ambassador to the United States. President Roosevelt was about to set up the National Defense Research Committee which would include a division devoted to radar.

The Tizard mission had two major achievements.

First, it facilitated the practice of designing and testing weapons by teams which brought together military officers and civilian scientists for regular discussions and trials. An unforeseen aspect of this cooperation was a strong sense of mutual understanding which strengthened the Western Alliance after America had entered the war. The mission had included among its members E G ("Taffy") Bowen who had been in the Air Ministry team, and John Cockcroft who had been brought into experiments with coastal radar stations as well as into the Ministry of Supply. They both stayed on after the mission had broken up in order to be on hand to give advice that came from direct practical experience of war. Bowen became much in demand among American experimenters. Tizard had gained the confidence of Alfred Loomis, a millionaire financier and amateur physicist who was also a cousin of Henry Stimson, President Roosevelt's Secretary of War. Loomis's wealth and personal connections enabled him to cut through the secrecy surrounding different military hierarchies.

Second, the Tizard mission gave the Americans an opportunity to use the cavity magnetron, the newly developed generator that facilitated the use of radio waves of a much shorter length. This invention was promoted by a valve development contract between the Admiralty and the physics laboratory of Birmingham University; the prototype was built and tested in February 1940. The first application of radar to anti-aircraft guns used waves of 5.5 metres. All available evidence showed that smaller centimetric waves were preferable because they enabled the use of more mobile equipment.

In response to these mission exchanges, the Americans set up the Radiation Laboratory at the Massachusetts Institute of Technology, and its team of scientists under Louis Ridenour began immediately to concentrate on the development of "gun-laying radar" for anti-aircraft guns. Their priority was to see if they could design a system that would allow radar to track an enemy aeroplane automatically by "locking on" to the target and give what in British parlance was called "unseen fire control". Such a system required stronger and narrower beams which could be provided by the cavity magnetron.

On the British side, the improvement of gun-laying radar seemed to slip down the list of priorities in defence research because those in charge of grand strategy thought it more important to discover the character of the radar being used by the enemy, so that they could determine the best methods of attack rather than improve the system of air defence.

The air intelligence section of the Secret Intelligence Service (M16) concentrated on acquiring an accurate knowledge of the Luftwaffe's radar. By relying on the decrypts of German messages intercepted by the code breakers in Bletchley Park, and on the analysis of air reconnaissance photographs at the interpretation centre in Medmenham, the principal scientist, R V (Reginald Victor) Jones, who lived in a flat in Richmond Hill Court and was thoroughly familiar with Petersham and its war-time role, was able to identify the radio beams used to guide German bombers to their targets and to judge the effectiveness of German radar surveys on British aircraft.[9]

At a meeting of Cabinet on 21 June 1940, Jones was able to demonstrate that the Germans possessed a system of night navigation which brought bombers to their targets by radio beams – code named "Knickebein" or "crooked leg". Winston Churchill, who agreed to share British secrets with the American authorities, was largely persuaded to give his permission after being convinced that the Germans already had effective defensive radar.

The study of enemy radar magnified the importance of securing American collaboration. Although the United States did not enter the war until December 1941, American "liaison officers" were sent to Britain as part of the Tizard mission exchanges.

[9] His memoirs, *Most Secret War* (1978), remain an authoritative account of the role of civilian scientists.

4: Anti-aircraft radar

PETERSHAM BECAME A MAJOR NODE in the network of civilian scientists advising the armed forces who had begun to experiment with "radio direction finding" (RDF). The term RDF was then abandoned in favour of the American formula "radio detection and ranging" (radar), a system using radio waves to detect the presence and location of distant and moving objects.

A radar dish or antenna transmitted a radio beam as a series of pulses, the echoes of which were then reflected back to a receiver from any object in its path. In the initial stages of experimentation the receiver needed protection from the much stronger power of the transmitter so that it could register the echoes as blips on a cathode ray tube. The length of the radio wave determined the width of the scanning beam. The two most important problems in radar development were narrowing the scope of the beam by shortening its wavelength and reducing the size of the equipment by combining transmission and reception in the same device.

Petersham became a major node in the network of civilian scientists advising the armed forces who had begun to experiment with radar

The improvement of anti-aircraft guns by using radar to direct their fire was a difficult task. It required a method of feeding the data acquired by radio waves into the predictors that gave the coordinates of objects moving in the sky and provided instructions on how the guns were to be aimed – direction, angle of elevation and distance. Both types of predictor then in use had been designed for plotting the position of the enemy by visual means in daylight or searchlight, not for spotting bombers flying at night.

The plotting crews were accustomed to giving instructions to the battery by word of mouth, announcing the range, the elevation, and the bearing of the guns, as well as the fuse setting of the shells and the moment to fire. The challenge was to find a way of linking the radar data to the predictors automatically. The most elusive objective was "unseen fire control" – a beam that would home in on and automatically follow any hostile object. That possibility seemed very remote. Anti-Aircraft Command did not enjoy a high prestige among artillery experts.

Lindemann, from the moment he joined the Tizard Committee, disparaged the application of radar to anti-aircraft defences. At that date he had advocated the use of curtains of small explosives hanging on cables that could be dropped by parachute in front of approaching bombers. As a statistician he calculated that the odds against an anti-aircraft shell hitting an enemy aeroplane were very high.

Later in the war Lindemann appears to have persuaded Churchill that rockets would be much more effective than anti-aircraft shells. Rocket batteries could also be manned by crews with much less specialist training. They provided what was called "salvo fire".

The greatest investment in rockets for use against aircraft took place in the projectile development establishment of Woolwich Arsenal that had been evacuated to Fort Halstead in Kent on the North Downs. Duncan Sandys, Churchill's son-in-law, was placed in command of an experimental unit using rockets. With this kind of patronage Anti-Aircraft Command was obliged to begin to apply these techniques.

Professor Hill's contribution at this stage was to advocate the establishment of groups of high-powered searchlights that would expose the presence of enemy aircraft. But after a number of tests he recognised the superior value of radio direction finding.

After a number of experiments with radio waves at Daventry and Orford Ness, the Air Ministry had acquired Bawdsey Manor in Suffolk as a research laboratory in which to design what became the "Chain Home Low", the line of radar transmitters and receivers placed at suitable points along the country's eastern seaboard in order to give advance warnings of enemy aircraft approaching. The scientists who were pioneers at Bawdsey became important actors in all the subsequent experimentations.

Robert Watson-Watt from the National Physical Laboratory at Slough, who had supervised the experiment undertaken by Arnold Wilkins in February 1935 to demonstrate the radio detection of aircraft, became a leading figure. He and A P Rowe, from the Air Ministry itself, led the team which moved from Bawdsey and was then named the Telecommunications Research Establishment (TRE), first at Dundee, then at Worth Matravers near Swanage in Dorset and finally at Malvern in Worcestershire.

The Army had a small but significant contingent in the Bawdsey Manor team. A network of personal contacts about radar was built on this experience.[10]

Bawdsey Manor (formerly RAF Bawdsey) in 1992 *Russ McLean/ Wikimedia Commons*

Evidence of Bawdsey Manor experimentation and lines of thinking was apparent in the operational research section of RAF Fighter Command at its Bentley Priory headquarters. In 1937-38 Tizard and Rowe had sent a civilian scientist, B G Dickins, to Biggin Hill to help RAF officers understand the effects that radar might have. A Canadian scientist from Bawdsey, Harold Lardner, brought to Anti-Aircraft Command Headquarters in Stanmore many of the lessons of "operations research" which had been learnt in various radar exercises, particularly those acquired in the summer of 1939 before war was declared. Lardner knew the difficulties staff might encounter in "filtering" radar signals. Further operational procedures using radar were devised by the scientists and the crews of fighter squadron no. 32 at Biggin Hill. Tizard himself took part in the trials.

Priority in radar development, not surprisingly, was given to the RAF's desire for a system of "controlled interception", for the simple reason that British manufacturers were never going to provide sufficient aircraft to equal the number that the Germans could deploy. The rearmament programme was not large enough to counter the Nazi menace.

From a British point of view the conduct of operations by Bomber and Fighter Commands was going to depend on the best use of the aircraft available; each machine had to be mobilised most economically, with

[10] The Bawdsey Manor Group (www.bawdseymanor.co.uk) nowadays seeks to retain public interest in the achievements of the 1930s, and the Purbeck Radar Museum Trust (www.purbeckradar.org.uk) celebrates the two years that the research team (later to become TRE) spent in Dorset (May 1940 to May 1942).

Having been requisitioned for occupation by TRE, Malvern College boasts that the Second World War was won on its grounds, not like the Battle of Waterloo on "the playing fields of Eton". The Malvern Museum of Local History (https://malvernmuseum.co.uk) has a special radar section, and a publication by Ernest Putley, *Science comes to Malvern* (2009).

minimum turn-around times and rapid repairs. The production of a radar screen that could be fitted into each pilot's cockpit made it possible to begin to plan for the identification of the enemy in the air and the control of interception from the ground.

The most important innovation in the use of radar for the RAF was "ground-controlled interception" (CGI) – first used successfully during the Battle of Britain on 18 October 1940. This linked several radar stations to a command centre that guided fighter aeroplanes onto their targets. Several connections of this kind were in operation by early 1941. Fighter Command was the first to learn from what came to be called "operational research".

Taking advantage of all the experimental efforts, the Air Council formally created separate operations research sections for each command in the RAF in October 1941, and established at the same time an Operations Research Committee which included representatives from Anti-Aircraft Command, the Ministry of Aircraft Production, and the Ministry of Supply. For discussions of radar development its most important sub-committee was that on tactical counter measures. From time to time the committee convened a conference that could discuss technical points. The scientific advisers attached to different ministries when they met informally after June 1942 tended to emphasise the importance of studying the use of existing weaponry in operations. They thought that getting the best scientists into operational research would be much more rewarding than placing them in research teams devoted to new apparatus.

The priorities for Bomber Command lay in the creation of navigational and bomb aiming devices. Progress was made in systems for radar mapping of the ground flown over and for air-to-surface-vessel detection (ASV).

The first sinking of a German submarine by using ASV was in January 1941; the first major bombing raid using navigational aids was over the Ruhr in March 1942. A centimetric wave device (codenamed H2S) to assist "blind navigation" developed in TRE with the assistance of Alan Blumlein (1903-1942), the inventor from EMI who had created the line-type pulse modulator, allowed bombers to penetrate even deeper into Germany. Blumlein was killed in an air crash when testing his own invention.

5: The purpose of the Petersham unit

THE BASIC PURPOSE OF ANTI-AIRCRAFT COMMAND in establishing the Petersham unit was to take advantage of the increasingly obvious achievements of civilian scientists in applied research. The Command had 12 divisions in Britain by the end of 1940, some hastily formed from Territorials and from the Royal Regiment of Artillery.

There were 280,000 men at arms in many different theatres of war, and 74,000 women from the Auxiliary Territorial Service (ATS) serving largely as range finders and plotters on the maps of each battery. Men and women began to serve together – an action that seemed daring when the War Office was forbidding the Home Guard to recruit females. A film called *The Gentle Sex* released in 1943 emphasised the contribution of women searchlight operators to the fight against enemy bombers.

The first "mixed batteries" were those in Richmond Park which Winston Churchill himself came to inspect in 1941.[11] Petersham was an integral part of the Command. The General Post Office engineers were instructed to lay special telephone lines connecting Petersham with the Anti-Aircraft Command headquarters in Stanmore and with other war rooms.

The whole Petersham exercise was about the presentation of science to the military

From his headquarters General Pile had observed how the Bawdsey Manor experiments had benefited Fighter Command operational procedures. He wanted to overcome the resistance of old-fashioned senior Army officers to new ideas, such as operational research. He used to say: "No really new idea could be accepted by the War Office, because if it was thought to be new, it must have been thought of already, and if it was not in use, it must have been rejected on very good grounds."

Pile agreed with the judgement of one of his scientific advisers: "the introduction of new ideas rests solely with a commander or at least with his immediate deputy... report writing is a poor substitute for a senior officer

[11] See Simon Fowler, *Richmond at War 1939-1945,* Richmond Local History Society, 2015.

Winston Churchill with his daughter Mary and General Sir Frederick Pile (GOC Anti-Aircraft Command) watch anti-aircraft guns in action against V1 flying bombs, 30 June 1944. *©IWM*

who can discuss an idea with his equals." There were dangers when "trafficking in ideas far above our rank". The whole Petersham exercise was about the presentation of science to the military.

A fundamental problem was the ineffectiveness of anti-aircraft guns. It was pointed out: "It isn't easy to shoot down a plane with an anti-aircraft gun... Instead of sitting still, the target is moving at anything up to 300 mph, with the ability to alter course left or right, up or down. If the target is flying high it may take 20 or 30 seconds for the shell to reach it, and the gun must be laid a corresponding distance ahead. Moreover, the range must be determined so that the fuse can be set and, above all, this must be done continuously so that the gun is always laid in the right direction. When you are ready to fire, the plane, though its engines sound immediately overhead, is actually two miles away. And to hit it with a shell at that great height the gunners may have to aim at a point two miles farther still. Then, if the raider does not alter course or height, as it naturally does when under fire, the climbing shell and the bomber will meet. In other words the raider, which is heard overhead at the Crystal Palace, is in fact at that moment over Dulwich; and the shell which is fired at the Crystal Palace must go to Parliament Square to hit it."[12]

The guns made a lot of noise – thought good for the improvement of civilian morale as evidence of "fighting back" – but often caused damage to property with fallen shell cases. The most elementary "operational research" showed that 6,000 shells had been fired for every enemy aircraft destroyed. Winston Churchill in his history of the war wrote: "This roaring cannonade did not do much harm to the enemy, but it gave enormous satisfaction to the population".[13] On the second night of the London Blitz, General Pile had deliberately ordered "every gun was to fire every possible round". He noted that the barrage "bucked people up tremendously".[14]

The principal move made by government to improve Anti-Aircraft Command's use of its guns was the promotion of the Air Defence Experimental Establishment (ADEE), which had recently moved from Biggin Hill to a house called Bure Homage at Christchurch in Hampshire. It became the Air Defence Research and Development Establishment (ADRDE). John Cockcroft was appointed as its director on his return in December 1940 from a long stay in the United States after the Tizard Mission.

Several members of the ADRDE staff had before the war been seconded to Bawdsey Manor from the Air Defence Experimental Establishment (ADEE) at Biggin Hill and the Signals Experimental Establishment at Woolwich in order to form a small Army contingent led by Edward Paris, who had then been appointed deputy director of scientific research in the War Office.

While at Bawdsey, Paris had been engaged in experiments with gun-laying radar. He therefore brought the Army into direct contact with the network of radar scientists based on the Telecommunications Research Establishment (TRE). His team in 1937 had experimented with a fire control system which fed radio readings into the predictors, using a beam of a six-metre wavelength to measure the range of an aircraft. Calculations of elevation and bearing were more difficult to include. This work led to the production of Gun Laying

[12] Quoted in *A Roof Over Britain, The Official History of the AA Defences*, p.6, HMSO, 1943. See also https://spartacus-educational.com/2WWantiaircraft.htm

[13] Winston Churchill *The Second World War*, Vol. II, p.30 (revised edition), 1950

[14] Juliet Gardiner *Wartime: Britain 1939-1945*, p.341, 2004

Mark I sets. Unfortunately, few of these were available during the Battle of Britain; some were modified by the addition of an elevation-finding attachment.

The ADRDE was part of the Ministry of Supply to which Edward Paris was then appointed as the controller of physical research and signals development. Its terms of reference included determining the design and production of radar sets for Anti-Aircraft Command. Its staff was responsible for developing Gun Laying Mark II sets which had a gearing system to transmit radio data and then Gun Laying Mark III sets, first tested in June 1941, which used much shorter radio waves.

The application of radar to each battery using Gun Laying Mark I sets required the range measurements to be translated onto the rotation of a spindle that fed the predictor. It was important to find ways of calibrating points on the spindle. The shells were timed to explode according to the length of the range. Another important innovation generally desired was to create a "proximity fuse" that would time the explosion more efficiently.

The Gun Laying Mark I radio direction finding sets were fixed to anti-aircraft guns for the first time in Raynes Park on the night of 19/20 September 1940.

6: The initial set up

THE ANTI-AIRCRAFT COMMAND ESTABLISHMENT AT PETERSHAM had two functions. First, it was a home for scientists who could apply the principles of "operational research" to the practices of gun batteries.[15] Second, it provided a training school in which officers could learn about the findings of that research and then be posted to a battery where the knowledge acquired was of value.

General Pile's wish to overcome resistance to innovation gave prestige to the second function. He knew that his battery crews were going to find radar sets very difficult to understand, particularly the elevation finding attachments that were supposed to replace the predictors which depended on telescope manipulation. The gun-laying manuals for radar were so secret that they were not easily available to the officers in charge. Each battery needed its own resident radar expert. The Petersham-trained men sent out to each battery were initially civilian advisers, but the authorities quickly realised that such postings should carry the authority of a commissioned officer. The trainees were therefore subsequently given the rank of Second Lieutenant, first in the Royal Army Ordnance Corps (RAOC) and then in the Corps of Royal Electrical and Mechanical Engineers (REME).

After the move to Petersham, "Blackett's circus" expanded in numbers. L H Matthews, a 39-year-old zoologist from Bristol University, went on to specialise in position-indicating systems for Pathfinder bombers. David Lack, a 30-year-old ornithologist who had been studying the robin and Darwin's finches, was able to identify migrating birds when they put confusing images on radar screens. Nevill Mott, the professor of theoretical physics at Bristol University, acted as consultant on several radio wave problems.

J S Hey, the radio astronomer who discovered the importance of radio waves from the sun, became the Petersham expert on the radar jamming device known as "window" – strips of paper with black and silver paper sides dropped from aircraft. Raymond Beverton, a zoologist aged just 20 in 1942, developed insulation for radar cables. Patrick Johnson, a bachelor don and

[15] Operational research is a discipline that deals with the development and application of analytical methods to improve decision-making.

physics tutor in Magdalen College, Oxford, started a military career for himself by joining the team.[16]

The university qualities of the research team were apparent in the steady stream of reports and papers which its members wrote and which in some instances were significant for the extension of knowledge after the war. The study of radio astronomy, for instance, benefited from Petersham research. The most immediate benefits for anti-aircraft batteries over the whole country were those studies that led to the relocation of guns and to the construction of platforms for the radar cabins.

The Petersham research team brought about a regrouping of the batteries around London in order to make better use of the few gun-laying radar sets then available. It demonstrated that the concept of "a complete cover" for London was an illusion based on daylight experience. The Luftwaffe was bombing areas at random at night, not precise targets. The limited numbers of radar sets were allocated to eight-gun batteries formed by joining two four-gun batteries together. The number of "rounds per bird" was reduced from 6,000 shells to 4,000, and later to an even lower rate.

The team also demonstrated that the elevation attachment worked more effectively if the radar cabin was surrounded by an artificial level platform of wire netting supported on stakes, from which radio waves would also reflect. The radio beam scraped the ground as well as reaching high into the sky. Anti-Aircraft Command bought 3,500 miles of stay-wiring. The adaptation of anti-aircraft batteries to this discovery brought about a national shortage of chicken wire for farmers and horticulturalists.

The research team was directly engaged in the "centimetric revolution" brought about by the use of the cavity magnetron. By the spring of 1941 this valve was capable of giving a power of 40kW to a 10cm radio beam. The shorter the wave length meant the smaller the dimensions of the antennae, such as those suitable for being mounted in aircraft.

Furthermore, the ADRDE had made a gun laying set that that used a single cabin instead of the two cabins needed for the transmitter and receiver in Gun Laying. Marks I and II. Its design for Gun Laying Mark III which used a

[16] The later careers of key members of Blackett's research team are listed in Appendix 1 (p.48).

10cm beam came close to the ideal of "unseen fire control". The team examined modifications in the design of the radar transmitters and receivers. The invention of a "transmit/receive switch" allowed a single antenna to perform both functions.

The Petersham research staff tested new equipment by using barrage balloons, by chartering aircraft to fly over the village and by placing batteries out in the fields.

Among the most regular visitors to Petersham were representatives of the manufacturing firms that held the contracts for production, especially T H Bedford of A C Cossor Ltd and J T C Milne of EMI. They took away new specification proposals and came back with prototype models for testing. Great improvements were made in the calibrating mechanisms that controlled the elevation of the gun. Bedford designed mechanical training apparatus which operators could use in the village hall.

The Petersham research staff tested new equipment by using barrage balloons and by chartering aircraft to fly over the village. On one occasion in 1943 an Avro Anson aeroplane being used crashed onto the roof of Elm Lodge. Fortunately, the school principal at that moment was not in his office; he had taken the afternoon off to go to a Richmond cinema.

The most effective testing of any innovation had to be not in the laboratory of the village hall or All Saints' but out in the fields where batteries were placed. The research team instigated the construction of mobile laboratories that could take observers from battery to battery. These "recording vans" carried a document recording camera which photographed the output dials on a dial bank connected to each battery. Some members of the team spent many hours simply taking note of "operations" to secure data objectively. In the autumn of 1941 these vans first developed in Petersham were a major contribution to operations research elsewhere. After D-Day in 1944 eleven of them were shipped to 21st Army Group.

In the radio school, Ratcliffe made clear from the beginning that the period of training was to be a six-week course at a university graduate level. The time spent in training later became more varied, between five and 12 weeks according to the trainee's previous experience. Anti-Aircraft Command arranged for lower-level courses aimed at technicians or operators to be given in technical colleges across the country, such as Portsmouth Municipal and Coventry Technical. The textbook, *Radar Simply Explained* by R W Hallows, published after the war in 1945, was the product of courses prepared for gun control operators in colleges of this kind.

But Petersham was seen as "the radar university". It was assumed that anyone attending had a first- or second-class honours degree. There were only 20 on the first course, all former schoolmasters; they were obliged to eat at The Fox & Duck, because the canteen in the basement of the village hall had not been completed. Thereafter the pattern was established of 60 new students per fortnight. Ratcliffe promised to train 480 by this method as soon as possible. There seems to be no evidence of the final total of the school's "graduates". Some of the schoolmasters, such as Edward Humphreys from High Pavement School, Nottingham, later went into the operational research side.

About 300 Americans passed through between 1941 and 1943, some of them arriving before Pearl Harbour as a consequence of the Tizard mission for scientific collaboration.

The village institute at Petersham, with the inscription "Thy Kingdom Come".
Andy Scott/Wikimedia Commons

The Canadian Signals Regiment took over Kingston Girls' School in order to collaborate with the Petersham training school. For their first visit two Canadian officers were given a map reference and a code phrase "Thy Kingdom Come"; it took them some time to discover the location because they did not see immediately that the code phrase was carved into a panel over the main door of the village institute.

The trainees after an introduction to the general principles of radar then specialised in either guns or searchlights. The application of radar to searchlights was still at an experimental stage.

The great majority of trainees were men. Women were not admitted until 1943. Marjorie Inkster has written a memoir which covers her time on the course.[17] She has recollections of the smell of the food in the canteen which was normally full of Woodbines cigarette smoke.

University qualities on the training school side were a direct result of government policy for the promotion of scientific manpower. Churchill on forming his coalition government in May 1940 invited the former Cabinet Secretary, Maurice Hankey, who had been a minister without portfolio in Neville Chamberlain's Cabinet, to become the Chancellor of the Duchy of Lancaster with "special duties". These functions came to include the chairmanship of the Technical Personnel Committee, which gave him access to the central register of scientists kept by the Ministry of Labour. There was also a committee on skilled radio personnel which claimed to have recruited at different levels some 4,000 people into radar.

Hankey, who saw the need to find suitable people for radio engineering, persuaded 19 universities to put on special short courses about radio; he secured the funding for what were called "Hankey bursaries" which paid the fees and the maintenance of those on these courses. Through the Ministry of Labour he also deferred the call-up of 350 graduate schoolmasters so that they could be directed into channels where their knowledge could be immediately useful to the war effort.

The Petersham training school benefited greatly from a steady recruitment of "Hankey" candidates. Until he was demoted by Churchill to the post of

[17] See Sources on p.45.

Paymaster-General in July 1941, Hankey made occasional visits to Petersham. It had become a national centre for radio engineering expertise.

Hankey had also brought in the Canadians. Twenty-two members of the Royal Canadian Corps of Signals were lent to Anti-Aircraft Command in December 1940. They all reported for training to Jack Ratcliffe in Petersham, and were then dispersed to different batteries, some even in the Orkney and Shetland islands. Another 180 Canadians volunteered for training, and the Canadian Radio Location Unit formed in January 1942 manned radar stations on the south coast.

The trainees had to handle malfunctioning problems that might vary from battery to battery. Each location had its own characteristics. The resident radar expert with Petersham training needed a proper knowledge of the basic principles – hence the emphasis on graduate entry. A lot depended on how well the research had modified the training. The determination to bring together research and teaching set the university tone.

7: Subsequent developments

THE RELATIONSHIP BETWEEN RESEARCH AND TEACHING became more overtly political after the summer of 1941 because nearly all the Petersham experimentation was tied closely to the placing of contracts for the manufacture in Britain of radar equipment by the Ministry of Supply. Comparable work on weapons procurement in Canada and the United States that had been stimulated by the Tizard mission had fewer complications and delays.

The value of "operational research" on the behaviour of AA gun battery personnel was not recognised quickly enough to be translated into decisions that guaranteed the supply of specific pieces of equipment. The status of Anti-Aircraft Command in the minds of the Chiefs of Staff and others advising Churchill seemed insufficient to secure strong support for this kind of weapons development. The Army Council began to turn its attention to broader questions of "operational research" for other forms of equipment used in managing an offensive, such as the deployment of tanks, the use of amphibious boats or the distribution of lorries. Petersham began to feel less like an extension of RAF Fighter Command interests and more an asset to improve the Army's influence over the Ministry of Supply. The War Office seemed nearer than Anti-Aircraft Command headquarters.

Various theories circulated in the minds of those involved to explain this trend. The evidence was difficult for them to interpret while they were caught up in the day-to-day pressures of testing innovations to existing equipment. Some thought that perhaps too many key decisions in the allocation of contracts were taken at weekly lunches in the Carlton Hotel of those nicknamed "the boilermen" – the procurement officers of the principal ministries handling supplies.

After the war a number of participants gave their own versions of events. General Pile wrote his own account in *Ack-Ack,* published by Harrap & Co in 1949. He implied that the War Cabinet had deliberately downgraded the needs of anti-aircraft artillery. Churchill seems to have favoured letting Anti-Aircraft Command make do with Canadian machines. Cockcroft attributed the delay in British manufacture to a false start in design and indecision in policy making. Some suspected that Frederick Lindemann had sufficient influence to downgrade investment in gun-laying radar; they thought that he might have persuaded Churchill in August 1943 to order the suspension of manufacturing contracts for British gun-laying radar sets. Professor Hill believed that the anti-aircraft rocket that was "the dearly beloved pet of Lindemann" had been "a most infernal waste of time, effort, manpower and material".

The most dramatic interpretations of high policy decisions were given by C P Snow in *Science and Government* (1961) which made play of the tensions between Lindemann and Tizard, stressing a difference of opinion on the morality of strategic bombing aimed at the civilian population.

Sir Basil Schonland, pictured here in 1957, took over the running of the Petersham unit in 1941.

Photo by Walter Stoneman, ©National Portrait Gallery, London

The Petersham unit seemed to change its character when the research side became part of the chain of command under the Ministry of Supply. This shift followed the transfer of both Blackett and Radcliffe to other duties. Air Marshal Joubert persuaded Blackett to apply his intelligence to the problems of Coastal Command which was beginning to play an important role in the Battle of the Atlantic against German U-boats by using air-to-surface-vessel radar (ASV). In March 1941, Blackett moved from studying gun-laying radar to examining the radar sets fitted in aircraft that attempted to track submarines. Lord Hankey had previously tried to persuade Cockcroft to set up an operational research unit for Coastal Command.

In August 1941, Radcliffe returned to the Telecommunications Research Establishment from which he had originally been taken in order to run the radio school. There was a short period (March to August 1941) during which Radcliffe ran both research and training. For this time John Cockcroft as head of ADRDE sent a 28-year-old mathematician, Maurice Wilkes, to take charge of research on radio waves while Leonard Bayliss remained responsible for the gun-laying mechanisms. To all intents and purposes Wilkes was seen as an operational research extension of the ADRDE. This temporary arrangement was then formalised.

In July 1941 the Air Defence Research Group in Petersham was officially recognised as Air Defence Research and Development Establishment (Operational Research Group) (ADRDE [ORG]) within the Ministry of Supply. This name gave the term "operational research" a wider circulation.

The new director (later given the title of "superintendent"), Basil Schonland, had been Cockcroft's deputy at ADRDE for a short while. He had the advantage of possessing recent and direct war experience in the Egyptian Western Desert where during the winter of 1940-41 he had led a radar group of the South African Corps of Signals. As an expert on electrical currents and a former director of research on geophysics at the University of Witwatersrand, he was an obvious candidate for the appointment. The transformation of Petersham after Blackett was symbolised by Schonland being given an army rank and then gaining promotion – Lieutenant-Colonel in 1941, Colonel in 1942, and Brigadier in 1944.

From this point onwards the radio school under Patrick Johnson and then J A Harrison proceeded as a separate AA entity while the research side began to be seen as something that might benefit the Army as a whole.[18] Operational Research officers were needed in all theatres of war, not just Britain.

But the close contact between research and training staff was not diminished. The most valuable collaboration between the research unit and the school was their joint participation in the Saturday morning seminars and "Sunday Soviets" which visitors with the appropriate security clearance and an interest in the technical problems could attend; they attracted a mixture of soldiers and scientists. General Pile himself often came.

The term "Sunday Soviet" was coined during the London Blitz when Air Marshal Sir Philip Joubert used his house in Bournemouth at the weekends. As the assistant chief of staff responsible for radio and radar in the RAF, he got into the habit of visiting the Telecommunications Research Establishment (TRE) nearby at Worth Matravers on Sundays for informal discussions. These meetings helped to build good relations between scientists and serving officers. Furthermore, Petersham acted as the meeting place of several committees. Men from the AA batteries came to the users' committee; designers and manufacturers to the weapons committee. A German spy might

[18] This may have been John Audley Harrison (1917-1992) whose entry in *Who's Who* is uninformative.

have had difficulty in interpreting conversations overheard in the pub after these meetings had finished.

For those in charge of grand strategy the most obvious evidence of collaboration between the research and training sides in Petersham lay in the preparations they both made for the airborne raid on Bruneval near Le Havre in France, which took place during the night of 27/28 February 1942. Air reconnaissance and information from French resistance fighters had drawn attention to the presence of German Würzburg radar set at Bruneval. R V Jones of Air Intelligence was closely involved. Paratroopers guarding RAF radar mechanics trained in Petersham were dropped by parachute in order to dismantle the set and bring its parts back to Britain by boat. The research staff at Petersham needed to know how this particular set had been adapted for use at night in guiding German fighter aircraft; the training school wanted practice in dismantlement.

The research unit and the school jointly participated in Saturday morning seminars and "Sunday Soviets" which visitors with the appropriate security clearance and an interest in the technical problems could attend; they attracted a mixture of soldiers and scientists.

The very success of this raid led to the evacuation from Dorset of the Air Defence Research and Development Establishment (ADRDE) and the Telecommunications Research Establishment (TRE), both of which worked closely with colleagues in Petersham. The General Staff feared that the Germans might retaliate for the Bruneval raid by landing their special forces at Swanage and Christchurch in order to capture British radar "secrets".[19]

Having failed to requisition Marlborough College for the purpose, the Air Ministry moved TRE to Malvern College in Worcestershire. The evacuation meant transferring 350 families, of which 150 had children who were obliged to change school. ADRDE was moved to Pale Manor RAF station and adjacent buildings also in Malvern. The town was declared a "closed area" and made

[19] Ernest Putley has another explanation: TRE was moved to prevent German intelligence discovering information about its work by using British techniques for the detection of enemy signals.

subject to compulsory billeting by order-in-council. Accommodation had to be found for 2,500 scientists and staff. These relocations imposed on the staff at Petersham much longer journeys whenever they wished to consult colleagues in person.

By the spring of 1943 it looked as if the attempt to produce a gun-laying radar set of British manufacture was unlikely to bring better results than the parallel trials in the United States and Canada. There had been four slightly different designs for these Mark II sets, all using a 10cm radio wavelength, but with separate transmitting and receiving dishes and no system for automatic tracking on the target. They proved helpful to those who had previously been dependent on the acoustical horns that assisted searchlight operators, but the gun predictors could not cope with the flow of data. In the winter of 1943-44 the research team found ways of replacing instructions given on the telephone between predictor and battery with an automatic electrical communication.

Hopes were placed in GL Mark III designed by ADRDE and in a similar model developed in Canada. The Canadian model was manufactured and distributed in Australia and South Africa, but it was totally outclassed by American products. Mark III did not go into full production until March 1943. Just as Mark II had been made technically obsolete by Mark III, so Mark III was overtaken by American machines before it even began to be distributed. The concentration of "air defence" research in Malvern after the summer of 1942 reduced the number of personal contacts between the staff at Petersham and the ADRDE/TRE combination.

The drive for perfection in ADRDE was weakened a little from late 1943 onwards when John Cockcroft moved into the development of atomic energy. As a member of the mission to the United States in November 1943 which was concerned with the use of radar in the Far East, he had been asked to consider a shift in his responsibilities. He went to Canada in July 1944 to take charge of the Montreal Laboratory and of the NRX heavy water reactor on Chalk River.

By February 1942 the Americans had perfected a system of anti-aircraft gun-laying radar that tracked enemy aircraft automatically, giving "unseen fire control". The American team in the Radiation Laboratory successfully designed and tested a set that "locked on" to its target. This achievement was based on an adjustment in the radio beam which placed it slightly off centre and then on a rotation that formed a cone. A target caught in the overlapping signals from two cones could be tracked automatically.

There was also another major innovation. An analogue computer developed in the Bell Laboratories made it possible to direct the guns electrically through a servo-mechanism. When tests on the prototype showed that bombers could be tracked automatically from a distance of 18 miles, the American Army in April 1942 began manufacturing a radar set called SCR 584 using these methods. The Servo-Mechanism Panel of the Ministry of Supply, on which Arthur Porter represented Petersham, never made a comparable breakthrough.

The British Anti-Aircraft Command in 1944, on setting up its operations in Kent against the V1 rockets aimed at London, depended on American equipment of this kind acquired under the Lend-Lease agreement. Its gun batteries tried to destroy as many of these weapons as possible while they were in flight over open ground. The only defence against the much faster V2 rockets was a radar device, code-named "Oswald", which identified the launching sites which could then be attacked from the air. Such weapons could not be brought down when in flight.

The operations against the V1 rocket depended on the American SCR 584 and on American proximity fuses which had been acquired largely on the initiative of John Cockcroft after Churchill had instructed that the production of British GL Mark III sets should cease. Using his American contacts Cockcroft had arranged for a firing trial of the SCR 584 on the Isle of Sheppey in October 1943; he discovered that this apparatus was compatible with British predictors. With General Pile's support he managed to persuade the War Office to place an order for 134 SCR 584s from the Americans just in time for them to be used against the flying bombs.

The Tizard scientific mission to the United States in 1940 had helped to create conditions for weapons development that had not been hampered by the day-to-day pressures of defending the country against the threats of enemy attack. There was time and space in North America to accelerate arms production on a large scale.

By mid-1943 the staff of the Radiation Laboratory numbered 4,000, including around 500 research scientists, many of whom were physicists who then moved on to the Manhattan Project to work at Los Alamos on the creation of an atomic bomb. A lot of distinguished scientists had at some stage passed through the private laboratory of Alfred Loomis at Tuxedo Park and met in its surroundings those British colleagues who had similar interests. Denis Robinson from the Telecommunications Research Establishment, for instance, was in close touch with Norman Ramsey from the Radiation Laboratory who had been a scientific attaché in London.

American doctoral students who had been at British universities often played key roles. Ivan Getting, who was deeply involved with negotiations on contracts for the SCR 58, had a D.Phil. in astrophysics from Oxford.

Petersham scientists showed that modifications of the antennae of British radar sets made it possible to track the V2 rocket as it descended; they could predict with reasonable accuracy where it would fall. But the general public was not given any advance warning because the authorities feared that such advice might cause stampedes to the entrances of air-raid shelters in panic.

There were some minor Petersham successes. Experiments conducted in Richmond Park demonstrated that screens of wire netting placed in front of the antennae of the British radar sets greatly improved the detection of a low-flying bomb, the V1. Signals reflected from the Manor Road gasometer and from Walls' pork pie factory in Acton disappeared from the radar screen!

Petersham scientists also showed that modifications of the antennae made it possible to track the V2 rocket as it descended; they could predict with reasonable accuracy where it would fall. But the general public was not given any advance warning because the authorities feared that such advice might cause stampedes to the entrances of air-raid shelters in panic.

8: The needs of the Army

THE ARMY MOVED FAIRLY SLOWLY TO SET UP OTHER UNITS to cover subjects worthy of operational research; it began to consider broader questions than anti-aircraft gun laying. Its reaction to the failure of its coastal batteries to fire on the German battleships, *Scharnhorst* and *Gneisenau*, which had passed unscathed through the Straits of Dover in February 1942, was to insist on taking over from the Air Ministry all the work on problems that arose when its radar sets were jammed, either by weather conditions or by enemy action. It set up a mobile laboratory on the cliffs of Dover and instructed J S Hey from Petersham to take charge of what was called "J-Watch".

During the course of 1942 the Army authorities began a series of negotiations which changed the terms of reference of ADRDE (ORG). By July 1942 it was reporting to the weapons development committee, not the anti-aircraft and civil defence committee, and to Sir Charles Darwin, the newly appointed War Office scientific adviser, who had been Director of the National Physical Laboratory at Teddington and then the manager of Anglo-American scientific cooperation in Washington DC.

After further discussions in December 1942, it was formally renamed as the Army Organisational Research Group (AORG) in February 1943. The headquarters of the Group was set up in Ibstock Place, Roehampton, not Petersham.

The Petersham establishment remained committed to anti-aircraft problems, but became simply section No.1 alongside nine other sections across the Army as a whole. These included sections devoted to coastal gunnery, signals, and infantry. Sections 9 and 10 devoted to time and motion studies and to the circulation of all the reports were at Whitehall Court in central London.

Omond Solandt was transferred in 1943 from Lulworth to Petersham to be "deputy superintendent" under Schonland as "superintendent". A Canadian who had trained in medicine and undertaken postgraduate research in Britain, he directed the South West London Blood Depot during the Blitz and then, with the rank of Colonel in the Canadian Army, worked on the behaviour of tank crews in action – a good example of an increasing interest in operational research. He became the superintendent after Schonland left for Europe.

Omond Solandt in about 1940
The University of Toronto Archives

The AORG remained jointly under the supervision of the Ministry of Supply and the War Office until the end of hostilities. For a short while during 1945 Solandt was the scientific adviser to Mountbatten in South-East Asia Command. The surviving private post-war correspondence of Schonland suggests that the former colleagues of Solandt regarded him as a somewhat pretentious person. Patrick Johnson in a letter of 1945 wrote that he was "more of a bore than ever". His other comments were about having "really capable and personable officers".

With the assault of Allied armies on Nazi-occupied Europe, first in Italy and then in France, some of the operations research scientists from Petersham were invited to advise generals in the field. Blackett had written two reports in 1941 and 1943 on the basic principles of operational research.[20] David Hill had already been posted to North Africa; Patrick Johnson joined the D-Day landing forces in Normandy. Basil Schonland himself in May 1944 was sent to be the scientific adviser of General Bernard Montgomery in 21st Army Group. When told that Schonland was to observe the progress of the war and give

[20] See Blackett, pp.169-198.

advice on operation efficiency, Montgomery replied: "I observe my own battles."[21]

Many Petersham researchers continued to concentrate on the problems of gun-laying radar. But instead of having direct feedback from observations in the field which they had organised, some found themselves under the constraints of an army bureaucracy which issued directives from a committee. From February 1943 onwards there was a "Control Group" that approved research programmes.

The "university atmosphere" of the early days was somewhat diminished. The Petersham establishment had become part of a line of command. There was a short period at the end of the war when its members of staff were divided between the Ministry of Supply and the War Office. But the Army soon took control and moved everybody to premises in West Byfleet where a newly fashioned Army Operational Research Group occupied the manor house and forty acres of ground.

[21] Papers and correspondence of Sir Basil Schonland, 86/63/1 Imperial War Museum

9: Petersham's legacy

THE HISTORY OF PETERSHAM'S CONTRIBUTION to the application of scientific method to military operations was initially only written under the provisions of the Official Secrets Act. Summaries placed on file exist in The National Archives. The surveillance of enemy radar brought all the activities at Petersham under rules of very strict secrecy. Interfering with or jamming enemy radio signals became part of "radio warfare" and the fundamentals of "electronic intelligence" or "elint", run after the war by the RAF Countermeasures Group and then by the Government Communications Headquarters (GCHQ).

When the special contributions of GCHQ at Bletchley Park in deciphering German messages on wireless traffic that used the Enigma machine became public knowledge in the 1970s, the work of Petersham did not attract a comparable amount of attention. The sad story of failing to secure "unseen fire control" may have contributed to this neglect. Cockcroft, for instance, also regretted the failure of British scientists to develop properly a proximity fuse that would time the explosion of an anti-aircraft shell to the moment when an enemy aeroplane was close by. Photoelectric fuses used on the anti-aircraft rockets were ineffective. Only the American laboratories managed to develop the radio proximity fuse. The Petersham story did not have the appeal of success.

Bletchley Park's special contribution in the Second World War became public knowledge in the 1970s, but Petersham's contribution to the war effort did not attract a comparable amount of attention

Furthermore, the presence of so many distinguished scientists in the village for the duration of the war left little physical evidence of their activity. The bits and pieces of wiring found in the vicarage garden did not arouse any keen archaeological enquiry. There was no obvious question demanding an answer.

The requisitioned buildings in Petersham were restored to their owners in April 1946. The vicar, the Rev R S Mills (1882-1971), who was the incumbent of St Peter's from 1929 to 1963, wrote in the March 1946 issue of *The Petersham Leaflet*: "the establishment of the Radio School here involved many of us in inconveniences and compelled us to reduce our parochial duties to a minimum. It is no longer secret that some of the most important experiments that led to discoveries that made British radar the extremely efficient instrument of war that it came to be, were carried out in Petersham Vicarage." This praise for local scientists was understandable but a little exaggerated, because it neglected to mention the complicated network of consultations across all the institutions involved, including the American.

The principal achievement of the Anti-Aircraft Command's efforts in Petersham was to incorporate into Army thinking the notion of operational research

The principal achievement of Anti-Aircraft Command's efforts in Petersham was to incorporate into Army thinking the notion of operational research. General Pile succeeded in getting "a really new idea accepted by the War Office". The recruitment of university research scientists to study the problems of anti-aircraft gun fire had the longer-term consequences of bringing various forms of system engineering into essential parts of strategic planning.

After the formation of a single Ministry of Defence in 1964, the operational research sections of the three services merged as the Defence Operational Analysis Establishment (DOAE) at West Byfleet in 1965. This was renamed the Defence Operational Analysis Centre (DOAC) in 1992 and then the Centre for Defence Analysis (CDA) within the Defence Evaluation and Research Agency (DERA) in 1995.

The privatisation of research announced in 1998 led to the latter losing most of its staff to a public-private body called QinetiQ. But the operational analysis side of the work was considered too important to be placed in the private sector. Since 2001 anyone seeking the heirs to the Petersham enterprise will find them in the Defence Science and Technology Laboratory on the former Admiralty site at Malvern which had been HMS *Duke* and which remains firmly in the public sector. That institution is the principal surviving monument to Petersham's wartime sacrifice.

Sources

The National Archives

- CAB 44/48, A history of Anti-Aircraft Command
- AVIA 22/2306 and 2308, Operational Research and Organisation by Dr J A Ratcliffe
- WO 291/887 and 1288, The origins of Operational Research in the Army, by Leonard Bayliss, and AORG Radar Section 1941-5
- WO 291/303, The Development of HA Unseen Fire Control 1940-45, with special reference to the work of AORG
- WO 291/462, AORG observations on the development of AA No.3 Mark I (GL IIIc) in AA Command May-June 1943
- AIR 20/2377, paper 28 (10 Jan. 1942), The AA gun defence of Great Britain

British Library Sound Archive

- Oral history interview with Conway Berners-Lee (1921-2019). A brilliant Cambridge maths graduate, Berners-Lee trained at Petersham before becoming a technical officer with anti-aircraft batteries in London and Norwich

Imperial War Museum

- Docs 3382 Papers and correspondence of Sir Basil Schonland
- Docs 13912 Papers of Captain Stephen Paull, US Army
- Docs 12888 Memoir of Marjorie Inkster

Biographies and memoirs

- *Oxford Dictionary of National Biography* (entries on Beverton, Blackett and Cockcroft)
- Biographical Memoirs of Fellows of the Royal Society (see list in Appendix 1 on p.48, and also entries on Canadians, such as Wilfred Bennett Lewis 1908-1987)
- Guy Hartcup and T E Allibone, *Cockcroft and the Atom,* Adam Hilger Ltd., Bristol, 1984
- R V Jones *Most Secret War: British Scientific Intelligence 1939–1945*, Hamish Hamilton, 1978

- Marjorie Inkster *Bow and Arrow War: from FANY to Radar in World War II,* Brewen Books, 2005
- Arthur Porter *So Many Hills to Climb,* Beckham Publishing Group, Silver Springs, USA, 2008
- Mary Jo Nye *Blackett: physics, war and politics in the 20th century,* Harvard University Press, 2004

General histories

- *A Roof Over Britain, The Official History of the AA Defences,* HMSO, 1943
- P M S Blackett *Studies of War: nuclear and conventional,* Oliver & Boyd, 1962
- Louis Brown *A Radar History of World War II: technical and military imperatives,* Institute of Physics Publishing, Bristol & Philadelphia, 1999
- Colin Dobinson *Anti-Aircraft Command: Britain's anti-aircraft defences of World War II,* Methuen, 2001
- David Edgerton, *Warfare State: Britain, 1920-1970,* Cambridge University Press, 2006; *Britain's War Machine: weapons, resources and experts in the Second World War,* Allen Lane, 2011
- Simon Fowler, *Richmond at War 1939-1945,* Richmond Local History Society, 2015
- G Neville Gadsby, "The Army Operational Research Establishment", *Operational Research Quarterly* Vol. 16(1), pp.5-18, March 1965
- Jack Gough *Watching the Skies: a history of ground radar for the air defence of the UK by the RAF, 1946-1975,* HMSO, 1993
- Damien Lewis *Raiders of the Shadows,* Quercus, 2019
- R F H Nalder *The History of British Army Signals in the Second World War,* Royal Signals Institution, 1953
- Stephen Phelps *The Tizard Mission: the top-secret operation that changed the course of World War II,* Westholme Publishing, Yardley, Pennsylvania, 2010
- Frederick Pile *Ack-Ack: Britain's defence against air attack during the Second World War,* Harrap, 1949
- Ernest Putley *Science comes to Malvern: TRE, a story of radar 1942-1953,* Aspect Design, 2009
- A P Rowe *One Story of Radar,* Cambridge University Press, 1948

- Tom Whipple *The Battle of the Beams: the secret science of radar that turned the tide of the Second World War,* Bantam, 2023

Articles

- K Brian Haley, "War & Peace: the first 25 years of OR in Great Britain", *Operations Research* 50(1), pp.82-88, 2002
- R A Forder, "OR in the UK Ministry of Defence: an overview", *Journal of the Operational Research Society* 55(4), pp.319-32, 2004
- Victor Lown and Paul Mitchell, "Arnold Wilkins: pioneer of British radar", *The Historian* No. 107, pp.15-17, 2010
- J F McCloskey, "The beginnings of OR: 1934-41", *Operations Research* 35(1), pp.134-52, 1987
- David Zimmerman, "Information and the Air Defence Revolution. 1917-40", *Journal of Strategic Studies* 27(2), pp.370-94, 2004

Appendix 1

The principal scientists working in Petersham 1940-1946

Name	Born/ died	Elected FRS	Subject discipline	Subsequent career
Leonard Bayliss	1900-1964		Physiology	University of Edinburgh
Raymond Beverton	1922-1995	1975	Biology	Government Fisheries Laboratory; Head of National Environmental Research Council
(Sir) Patrick Blackett	1897-1974	1933	Physics	Nobel prize winner; Professor, Manchester University; Professor, Imperial College London
(Sir) John Cockcroft	1897-1967	1936	Physics	Director, Harwell Atomic Energy; Master of Churchill College, Cambridge
J S (Stanley) Hey	1909-2000	1978	Physics	Head of AORG; Head of research group, Royal Radar Establishment
David Hill	1915-2002	1972	Physiology	Marine biologist and muscle expert
(Sir) Andrew Huxley	1917-2012	1955	Physiology	Nobel prize-winner; Master of Trinity College, Cambridge
Patrick Johnson	1904-1996		Physics	RAF Cranwell; Scientific Adviser to Army Council and to SHAPE
David Lack	1927-1973	1951	Ornithology	Fellow of Trinity College, Cambridge
L H Matthews	1901-1986	1954	Zoology	Scientific Director of Zoological Society London
(Sir) Nevill Mott	1905-1996	1971	Physics	Cavendish professor, Master of Gonvile and Caius, Cambridge
F N R Nabarro	1916-2006	1971	Physics	Professor of Physics, University of Witwatersrand, South Africa
Arthur Porter	1910-2010		Mathematics	Instrument technology, servo-mechanisms in the United States
John A Radcliffe	1902-1987	1951	Physics	Director of Radio & Space Research Station
(Sir) Basil Schonland	1896-1972	1938	Physics	Chancellor of Rhodes University: Director of research, Harwell
Omond Solandt	1909-1993		Medicine	Director General of Defence Research, Canada
(Sir) Michael Swann	1920-1990	1962	Biology	Vice-Chancellor, University of Edinburgh
(Sir) Maurice Wilkes	1913-2010	1956	Mathematics	Built EDSAC computer in Cambridge: programming expert

Appendix 2

Petersham as a birthplace of radio astronomy

by Timothy M M Baker, 2024

Stanley Hey (1909-2000). *National Radio Astronomy Observatory/Associated Universities, Inc. Archives, Papers of Woodruff T. Sullivan III, Hey file*

STANLEY HEY, previously a Lancashire schoolmaster, became ADRDE's specialist on enemy radar jamming in 1942. In the course of his work at Petersham he discovered radio waves emitted by the sun and went on at Roehampton and Richmond Park to pioneer radio astronomy.

Hey's "J-Watch" ("J" stood for "Jerry") was established immediately after heavy jamming played a major part in the escape of German heavy ships during the "Channel Dash" of February 1942. It received reports from radar stations along the coast, of jamming, either by atmospheric conditions or by enemy action.

Only two weeks later, Hey received at Petersham reports from across the country of heavy jamming. By plotting how the source moved, he showed, despite widespread scepticism by his scientific colleagues, that the interference was coming from a large sunspot in the centre of the solar disc. This was the first discovery of solar radio emissions, and indeed the first discovery of a specific radio astronomical object. The finding was kept secret during the war for fear the enemy might schedule attacks to coincide with solar activity.

Together with Hey's later observations for AORG, when tracking V1 and V2 rockets from Richmond Park and Roehampton, of radar reflections of meteor trails, and of interference from cosmic radio sources, this work laid the foundations for the new science of radio astronomy, in which, immediately after the war, Hey's group became the British pioneers.

From 1945 Hey's modified radar equipment on the polo field at Sheen Cross in Richmond Park effectively became the first British radio observatory.

It directly influenced the work of later radio observatories in both Britain and Australia. Though two Americans had observed cosmic radio noise in the 1930s, Hey's identification in 1942 at the Petersham Old Vicarage of the radio sun thus has a strong claim to mark the birth of radio astronomy as a connected discipline.

The receiver on the Polo Ground in Richmond Park, with which Hey's group mapped the radio sky and discovered Cygnus A in 1946.
National Radio Astronomy Observatory/Associated Universities, Inc. Archives, Papers of Woodruff T. Sullivan III, Cosmic Noise

A detailed account of the work of Hey and his group at Petersham, Richmond Park, and Roehampton is in *Richmond History* 42 (2021/22), pp.22-27 and the Society for the History of Astronomy's *Antiquarian Astronomer*, 15 (2021), pp.2-14.

Timothy Baker is an amateur local historian and historian of science following a career in financial services policy in the City of London. He wrote London: Rebuilding the City after the Great Fire *(Phillimore, 2000).*